Toshimasa Yamazaki

Silent Speech Brain-Computer Interface in Japanese

AF153822

Toshimasa Yamazaki

Silent Speech Brain-Computer Interface in Japanese

LAP LAMBERT Academic Publishing

Imprint

Any brand names and product names mentioned in this book are subject to trademark, brand or patent protection and are trademarks or registered trademarks of their respective holders. The use of brand names, product names, common names, trade names, product descriptions etc. even without a particular marking in this work is in no way to be construed to mean that such names may be regarded as unrestricted in respect of trademark and brand protection legislation and could thus be used by anyone.

Cover image: www.ingimage.com

Publisher:
LAP LAMBERT Academic Publishing
is a trademark of
Dodo Books Indian Ocean Ltd. and OmniScriptum S.R.L publishing group

120 High Road, East Finchley, London, N2 9ED, United Kingdom
Str. Armeneasca 28/1, office 1, Chisinau MD-2012, Republic of Moldova, Europe
Managing Directors: Ieva Konstantinova, Victoria Ursu
info@omniscriptum.com

Printed at: see last page
ISBN: 978-3-659-60667-0

Dedication and Acknowledgements

My thanks and appreciation to Ken-ichi Kamijo (NEC Medical Solutions Division) for discussing about early plan for SSBCIJ.

I am grateful to many colleagues and students. Firstly, Maiko Sakamoto (Hitachi Public System Service Co. Ltd.), Kazufumi Tanaka (CARROLsystem, Ltd.), Minami Ouda (Software Vision Co., Ltd.), Takahiro Shibata (Olympus Software Technology Corporation), Shino Takata (Lincrea Corporation) and Hiromi Yamaguchi (NEC Information and Knowledge Labs.) were engaged in our early BCIs with motor imagery. Especially, I need to express my gratitude and deep appreciation to Kei-ichi Yamamoto (Japan Tobacco Inc.), Keiji Matsunaga (Dai Nippon Printing Co., Ltd.), Hiroshi Takayanagi (Future University Hakodate, Visiting Prof.) and Rika Kan (NTT DATA KYUSHU Corporation) who embodied my early idea for the JSSBCI. This result led me to get a Grants-in-Aid for Scientific Research on Scientific Research. Shuhei Ueno (SCSK Corporation), Yukito Hisano (Miharu Communications Inc.) and Masaya Yamada (INTEC Inc.) supported this project by generalization to three-dimensional formant frequency space, improved Kalman filter algorithm and grand averagings of actual and silent speech trials, respectively. Hiromi Yamaguchi and Ayaka Yamaguchi (Hitachi Systems, Ltd.) succeeded in silent season recognition by HMMs with formant frequencies and mel frequency cepstral coefficients (MFCCs). These results yielded one more Grants-in-Aid for Scientific Research on Scientific Research. Takashi Itoh (Kyushu Institute of Technology) first revealed the accuracy for all the seasons. Shun Hirose (Kyushu Institute of Technology) has constructed one-hiragana-character-HMMs (1-HC-HMMs) for the SSBCIJ, which might lead us to generalization to all the hiraganas. Takuro Shouno yielded the first results on the comparisons of Kalman and particle filters (Tokyo National College of Technology). Kyomoto Matsushita (Kyushu Institute of Technology) has been challenging a new approach to the SSBCIJ.

Prof. Takahiro Yamanoi (Hokkai Gakuen University) gave me EEG measurement system at early stage of my research. His attendance at international conferences had supported many students.

ECDL largely contributed to my research, and the ECDL was carried out by a PC-based software called SynaCenterPro which had been developed Ken-ichi Kamijo, Tetsuji Tanigawa (NEC Management Information Systems Division), Akihisa Kenmochi (NEC Medical Solutions Division) and Tomoharu Kiyuna (NEC Medical Solutions Division).

I am grateful too for the EEG measurement support and advice from Miyuki Giken Co., LTD, especially Atsushi Shirasawa, Eisaku Ohyama and Tsukasa Kajiwara.

Finally, my thanks must go also to Shin-ichi Fukuzumi (NEC Information and Knowledge Labs.) for the NEC financial support.

Abstract

Silent Speech Brain-Computer Interface in Japanese

Silent Speech Brain-Computer Interface in Japanese (SSBCIJ) is one of Silent Speech Recognition Systems (SSRSs), and the SSBCIJ enables us to decode silent speech in Japanese from single-trial electroencephalograms (EEGs). I believe that this BCI is the next generation and essential one, because the BCI would be the most natural and easy-to-do for users, differently from the previous BCI with motor imagery. In order to develop the JSSBCI, my research group carried out two experiments (Experiments I and II). Experiments I and II commonly consisted learning and decoding phases. In the learning phase, 19-ch EEGs and speech signals were simultaneously recorded during the actual janken (Experiment I) and actual season (Experiment II) tasks. Independent component analysis (ICA) then equivalent current dipole source localization (ECDL) were applied to the recorded EEGs. The relationship between spectrograms of the speech signals and the ICs whose dipole solutions were localized mainly to the Broca's area was described by Kalman filters (KFs). In the decoding phase, only the EEGs were measured during the silent janken (Experiment I) and silent season (Experiment II) tasks. By inputting the ICs whose dipole solutions were located at the Broca's area to the learned KFs, *silent* spectrograms were estimated. Moreover, in Experiment II, by inputting the estimated *silent* spectrograms to hidden Markov models (HMMs), log-likelihoods were calculated for each silent season task. From plotting of the estimated spectrograms in the F1 (the first formant frequency)-F2 (the second one) plane and the maximal log-likelihood, which janken and season, respectively, were silently spoken were predicted. In Experiment I, the confusion matrix with diagonal components of "correct" and other ones of "incorrect" was obtained for all the ten healthy subjects. In Experiment II, the accuracy was 91%, 89%, 75% and 100% for the silent "spring", "summer", "autumn" and "winter", respectively, only in a few of the subjects. This book addressed itself to the first step of the SSBCIJ. Finally, I discussed further expansions of the present SSBCIJ from various points of view.

Table of Contents

Chapter 1 Introduction: Significance of Silent Speech Brain-Computer Interface in Japanese (JSSBCI) as Silent Speech Recognition Systems (SSRSs)

Silent Speech Brain-Computer Interface in Japanese (SSBCIJ) is one of Silent Speech Recognition Systems (SSRSs) using scalp-recorded electroencephalograms (EEGs). The SSRSs are ones enabling speech communication to take place when an audible acoustic signal is unavailable, and are roughly classified into "physical" and "electrical" (Denby et al., 2010). The former technologies refer to the following four:

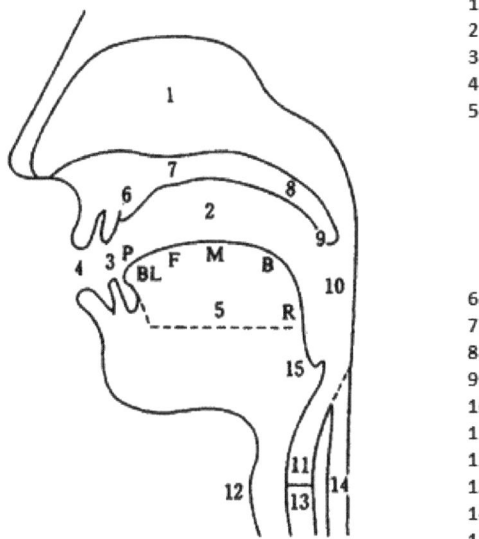

1: nasal cavity
2: oral cavity
3: tooth
4: lips
5: tongue
  P: lingual apex
  BL: tip of the tongue
  F: front lingual surface
  M: middle lingual surface
  B: back lingual surface
  R: root of the tongue
6: gums
7: soft palate
8: hard palate
9: uvula
10: pharynx
11: vocal cords / glottis
12: larynx
13: trachea
14: esophagus
15: epigottis

Fig.1-1: Vocal organs. Reproduced from Kubozono (2010) with permission.

(1) EMA: As the shaping of the vocal tract (Fig.1-1) is a vital part of speech production, a direct and attractive approach to SSRSs would be to monitor the movement of a set of fixed points within the vocal tract.

Fagan et al. (2008) developed an EMA (electromagnetic articulography)-based SSRS consisting of permanent magnets attached at a set of points in the vocal apparatus, coupled with magnetic sensors positioned around the user's head.

(2) US: Ultrasound (US) imagery is a non-invasive and clinically safe procedure which makes possible the real-time visualization of one of the most important articulators of the speech production system – the tongue. Hueber et al. (2010) constructed a large audio-visual unit dictionary which associates a visual realization with an acoustic one for each diphone.

(3) NAM: Non-audible murmur (NAM) is the term given to the low amplitude sounds generated by laryngeal airflow noise and its resonance in the vocal tract. The idea of applying NAM for telecommunication purposes was first proposed by Nakajima et al. (2003).

(4) Electromagnetic or vibration sensors: DARPA Advanced Speech Encoding Pilot Speech corpus (Tardelli, 2003), Defense Research and Development Canada (Bos and Tack, 2003) and the EU project SAFIR (Speech Automatic Friendly Interface Research, IST-2002-507427) (Dekens et al., 2008) had developed non-acoustic sensors for low bit rate speech encoding. For the purpose, glottal waveforms were investigated which can be used for de-noising by correlation with the acoustic signal obtained from a standard close-talk microphone. The waveforms may be obtained either via detectors which are directly sensitive to vibrations (e.g., Tardelli, 2003) transmitted through tissue – throat microphones and the like – or from the interaction of glottal movement with an imposed electromagnetic field (e.g., Bos and Tack, 2003).

The latter ones are as follows:

(5) sEMG: As speeches are produced by the activity of human articulatory muscles, the resulting surface electromyograms (sEMGs) measured at these muscles provides a means of recovering the speech corresponding to it. Such sEMG-based SSRSs allow for acoustic units smaller than words or phrases, enabling large vocabulary recognition systems (e.g., Schultz and Wand, 2010).

(6) Non-invasive EEG, MEG, fMRI and NIRS: Suppes et al. (1997) were the first to show that isolated words can be recognized based on EEG

and MEG (magnetoencephalography) recordings. Using a BCI usually requires the users to explicitly manipulate their brain activity, which is then transformed into a control signal for the device. This typically involves a learning process which may last several months (e.g., Neuper et al., 2003). In order to circumvent this time consuming learning process, the SSRSs have increasingly proposed using non-invasively recorded brain activity, such as scalp-recorded EEGs (Callan et al., 2000; Wester, 2006; DaSalla et al., 2009; Matsumoto and Hori, 2013; Riaz et al., 2013), functional magnetic resonance imaging (fMRI) (Naci et al., 2013) and functional near infrared spectroscopy (fNIRS) (Falk et al., 2013). However, these had been still in the experiment stage, and limited almost to vowel recognition.

(7) Invasive SSRS: Attempts have also recently been made to utilize intracortical microelectrode technology and neural decoding techniques to build an SSRS which can restore speech communication to paralyzed individuals (Guenther et al., 2006; Guenther et al., 2009; Brumberg et al., 2010), or to restore written communication through mouse cursor control BCIs for use with virtual keyboards (Kennedy et al., 2000) and the control of prosthetic hand with multiple joints (Hochberg et al., 2006).

These seven technologies are compared from various viewpoints as shown in Fig.1-2. Each axis is as follows:

· Works in silence – Can the device be operated silently?

· Works in noise – Is the operation of the device affected by background noise?

· Works for laryngectomy – Can the device by used by post-laryngectomy patients? It may be useful for other pathologies as well, but laryngectomy is used as a baseline.

· Non-invasive – Can the device be used in natural fashion, without uncomfortable or unsightly wires, electrodes, etc?

· Ready for market – Is the device close to being marketed commercially?

This axis also takes into the account in a natural way the current technological advancement of the technique, responding, in essence, to the question, "How well is this technology working as of today?".

· Low cost – Can the final product be low cost? The answer will depend, among other factors, on whether any "exotic" technologies or procedures are required to make the device function.

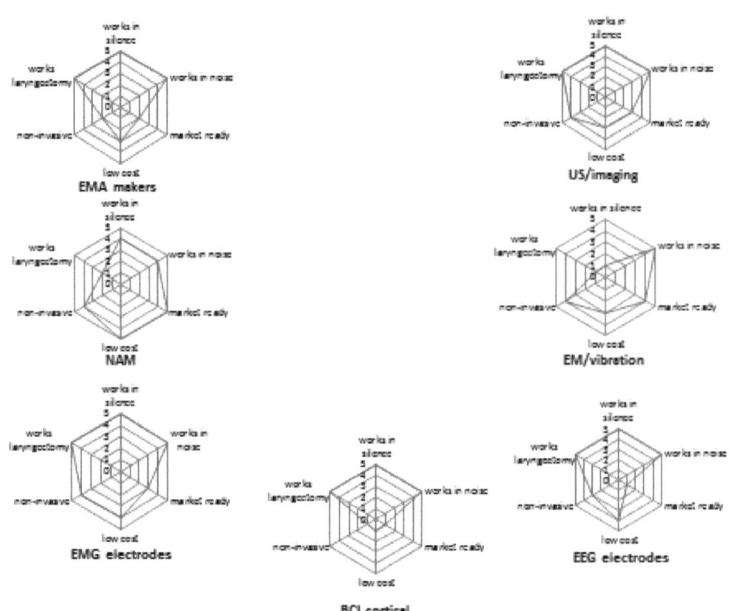

Fig.1-2: "Spiderweb" plots of the 7 SSRS technologies. Axes are described in the book. Reproduced from Denby et al. (2010) with permission.

Thus, I have greatly expected the "EEG electrodes". The reason is that my research group had succeeded in single-trial-EEG-based BCIs with motor imagery (Fig.1-3) (Yamazaki et al., 2013; Yamazaki et al., 2014; Sakamoto et al., 2015). For these BCI algorithms, applications of independent component analysis (ICA) and equivalent current dipole source localization (ECDL) to scalp-recorded EEGs are essential.

Single-trial EEGs treated in this book obviously have noises such as line noises, broad high-frequency spectra and EMG contamination (Makeig et al., 1996). Because signals and noises should be separated, the ICA will enable this separation. And, extracted signals as ICs will be labeled by the ECDL. These two technologies will be later described in details. Moreover, in future, if one could use dry and gel parallel type electrodes (Fig.1-4) (Parthenon Electrode, Miyuki Giken Co., LTD., Japan), "Non-invasive" in Fig.1-2 would be greatly improved.

Based on the ICA and the ECDL, my colleagues will address themselves to the SSBCI in this book. The outlines of our SSBCIJ algorithm, consisting two phases ("learning" and "decoding"), are as follows:

(1) Multi-channel EEG recordings during overt (learning phase) or covert (decoding phase) speaking of words in Japanese

(2) Applying ICA to each single-trial EEG and extracting ICs whose ECDL solutions were localized mainly to the Broca's area

(3) Only in the learning phase, simultaneous measurement of speech signals by a microphone, then transformation of the signals into spectrograms or mel frequency cepstral coefficients (MFCCs)

(4) Description by Kalman filters (KFs) of the relationship between the extracted ICs and the spectrograms or the MFCCs

(5) Inputting results at the above step (2) to the KFs, and estimation of *silent* spectrograms or MFCCs as outputs of the KFs

(6) Decision of which word was silently spoken, using plotting of the silent spectrograms in F1 (the first formant frequency)-F2 (the second one) plane and log-likelihoods by hidden Markov models (HMMs).

Here, in addition to silent vowel recognition, we will challenge silent word one including consonants. For the purpose, two Experiments I and II were designated, which test the silent vowel recognition and the silent word one, respectively.

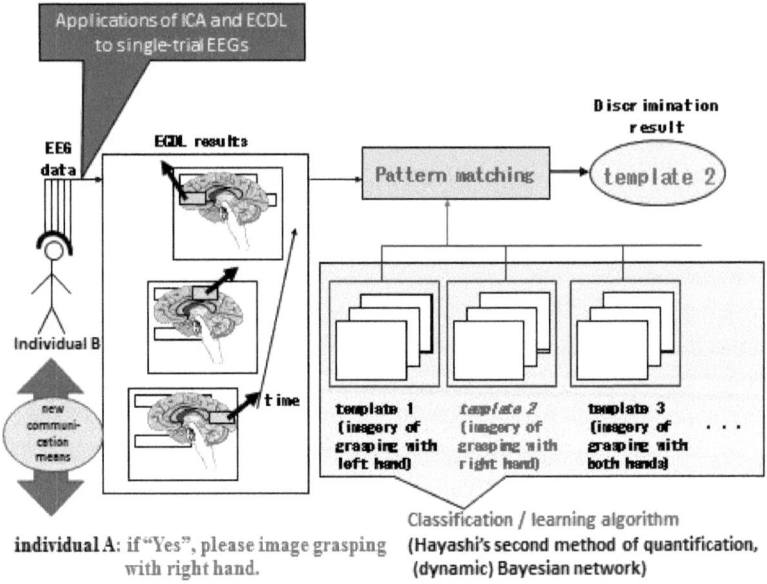

Fig.1-3: Our early plan for motor imagery BCI.

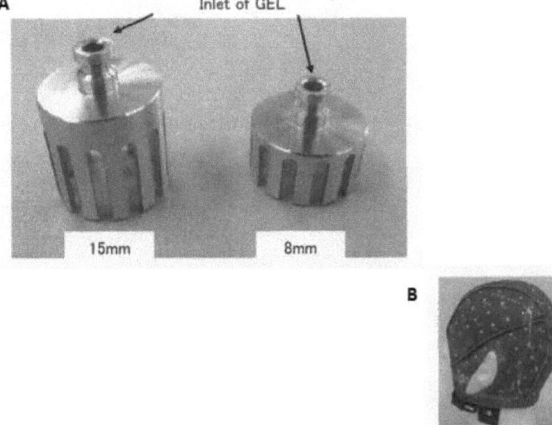

Fig.1-3: (A) Dry and gel parallel type electrodes; (B) electrocap which the electrodes are attached to.

Chapter 2 Experiment I: Silent Vowel Recognition

2.1 Materials and Methods

2.1.1 Subjects

Ten healthy student volunteers (two females and eight males; mean age: 23.7 ± 1.42 years) participated in Experiment I, whose procedures were approved by the Ethics Committee for Human Subject Research, Faculty of Computer Science and Systems Engineering, Kyushu Institute of Technology. Informed consents were obtained from all the students in writing for the procedures prior to the experiment. All the subjects were right-handed according to the Edinburgh inventory (Oldfield, 1971).

2.1.2 Tasks and EEG and speech signal data acquisition

The subjects were requested to speak "rock", "paper" or "scissors" (/guː/, /paː/ or /tʃɔki/ in English pronunciation of Japanese, respectively) into a microphone (MS-STM87SV, ELECOM CO., LTD., Japan) in the learning phase, or to silently speak it in the decoding phase, according to visual cues. After the subjects gazed for 3 s a point presented at the center of a monitor 62 cm away from the subjects, a line drawing of a hand indicating "rock", "paper" or "scissors" was presented for the next 3 s. Only the fixation point was presented for the next 3 s. Then, when the point disappeared, the subjects overtly or covertly spoke "rock", "paper" or "scissors" corresponding to the line drawing presented just before (Fig.2-1). The line drawings were randomly presented ten times for each janken. Nineteen active electrodes (AP-C100-0155, DIGITEX LAB. CO., LTD., Japan) were affixed to the scalp according to the International 10-20 system. Additive six channels were included for electromyograms (EMGs) and electrooculograms (EOGs), so that face, mouth and eye movements were monitored. The EEGs recorded at each electrode were fed to a amplifier (Polymate AP1132, DIGITEX LAB. CO., LTD., Japan) with 10000 gain and a notch filter of 60 Hz. The amplified EEGs were sampled at a rate of 1 kHz during an epoch of 3 s preceding and 3 s following each stimulus presentation. The on-line A/D converted EEG data was immediately stored on a hard disk in a personal computer (Fig.2-2). Note that speech signals collected by the microphone were digitalized and, if necessary, downsampled by Audacity (a free software for recording and editing sounds: http://audacity.sourceforge.net/), and transformed into spectrograms by WaveSurfer (a free audio & video software: http://www.speech.kth.se/wavesurfer/).

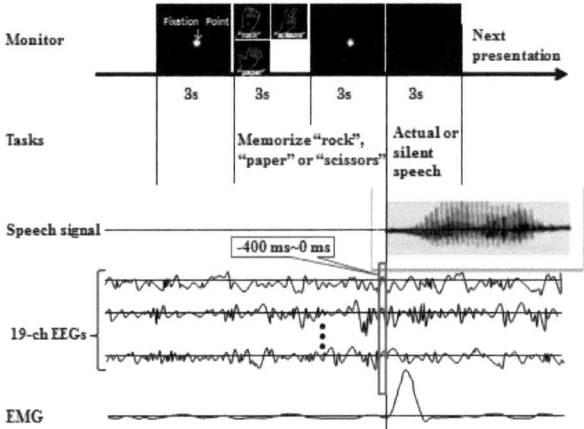

Fig.2-1: Task paradigm and time-scheduling of speech signal, EEG and EMG recordings.

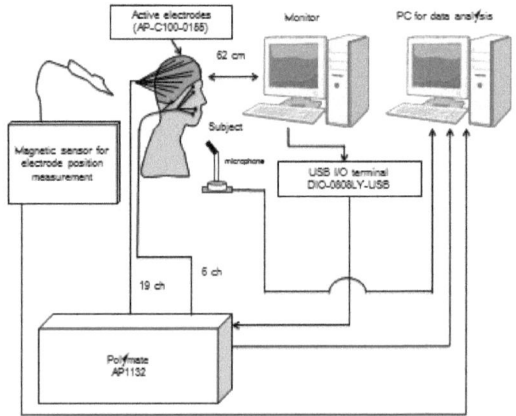

Fig.2-2: Experimental system for the measurements of EEGs, EOGs, EMGs and electrode positions and for stimulus presentation.

2.1.3 Data processing

In the learning phase (Fig.2-3), independent component analysis (ICA) was applied to the single trial EEGs obtained. ICALAB: http. www. bsp. brain. riken. jp/ICALAB/ICALABSignalProc/; was used to apply the fast fixed-point ICA algorithm (Hyvärinen and Oja, 1997) to the 19-ch EEGs, together with a MATLAB toolbox. Then, independent components (ICs) were extracted so that their equivalent current dipole source localization (ECDL) solutions were localized to the primary motor and premotor cortices, supplementary motor area (SMA) and/or Broca's area (BA) (Fig.2-4), with reference to the previous neuroimaging studies during overt articulation (Shuster and Lemieu, 2005; Vigneau et al., 2006; Peeva et al., 2010; Kielar et al., 2011; Croft et al., 2013) related to speech production.

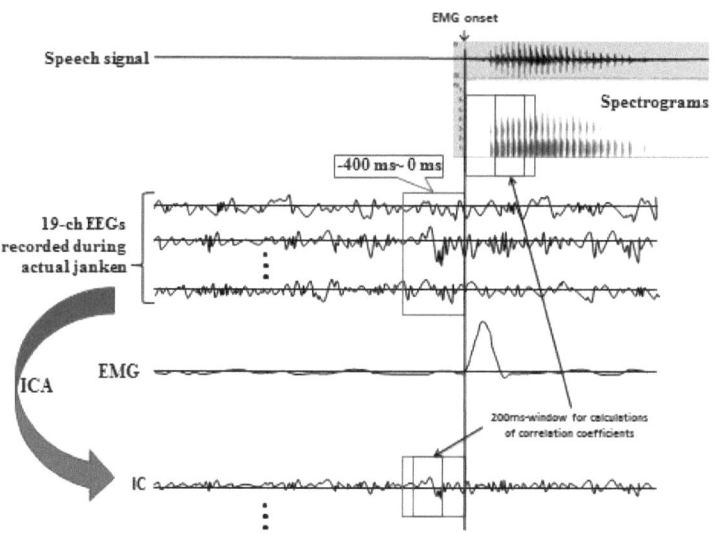

Fig.2-3: Learning phase in our SSBCIJ.

2.1.3.1 Grand averages

The grand average for the actual janken was obtained by the summation time-locked to the EMG onsets of the speech signals, and one for the silent tasks time-locked to average EMG onset of the signals for each task and subject. About 900 epochs were used for these grand averages, because three sessions were carried out, each of which included 10 trials for each janken. In the latter task, the EEGs were eliminated from the averaging if the subjects overtly spoke by mistake.

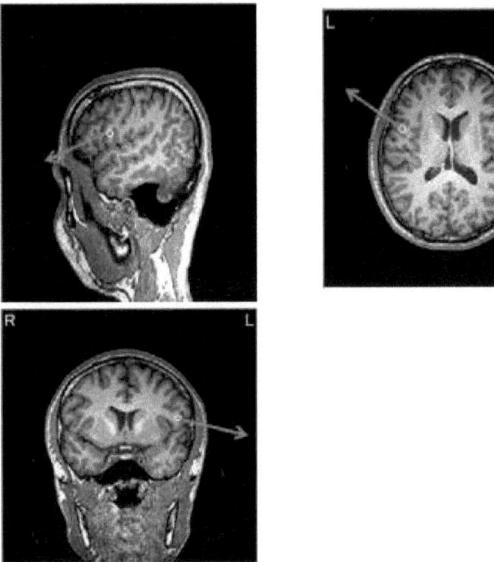

Fig.2-4: A representative ECDL result for one IC extracted from one single trial EEGs, where one dipole (blue arrow) was localized to the Broca's area.

2.1.3.2 ICA and ECDL

ICA refers to methods for obtaining statistically independent components from multi-channel-recorded data with time course. The ICA was originally proposed for solving blind source separation problem in speech recognition (Jung et al., 2001), which corresponds to cocktail party effect in cognitive psychology (Fig.2-5).

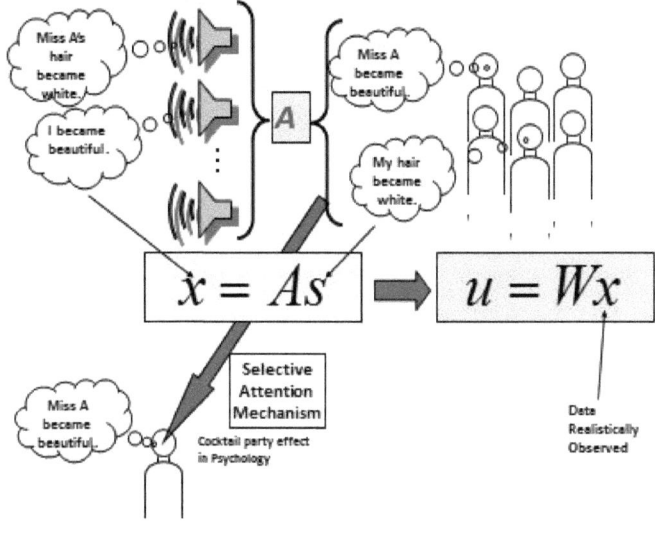

Fig.2-5: Independent component analysis for solving blind source separation problem.

Independent EEG sources obtained by ICA are dipolar (Delorme et al., 2012). ECDL was applied to the reconstructed EEGs, namely the projection of each of the ICs on the scalp surface by the deflation procedure, using "SynaCenterPro" (PC-based commercial software for multiple ECDL) (NEC corporation) (Fig.2-6).

The ECDL is a method for estimating neural generators as current dipole models in physics from multi-channel-recorded EEG data (Fig.2-7). The electrical activity of the brain consists of ionic currents generated by biochemical sources at the cellular level. These ionic currents cause electric and magnetic fields that can be measured in the brain and surrounding tissues. The behavior of these fields can be predicted because they obey physical laws. It is convenient to consider the generation of EEG signals in biophysical terms because this is the exact way to determine the potential distribution at the scalp given a set of intracerebral current sources, i.e., the so-called forward problem of electroencephalography. An understanding of this problem is necessary to discuss the inverse problem that constitutes the main concern of clinical electroencephalography, which is to determine the intracerebral sources given a measured potential distribution at the scalp. The inverse problem has no unique solution, as shown long ago by Helmholtz (1853); therefore, it is essential to understand the forward problem so that constraints of the inverse problem may be well examined (Niedermeyer and Lopes da Silva, 1987).

SynaCenterPro

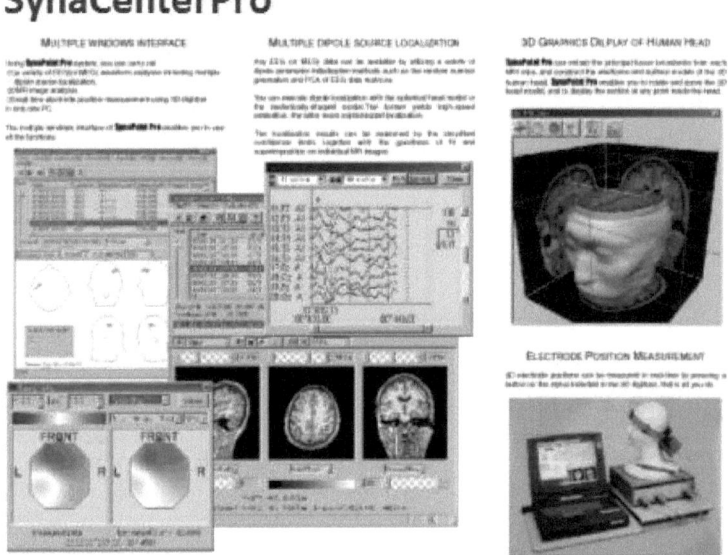

Fig.2-6: SynaCenterPro for ECDL.

"SynaCenterPro" software estimates unconstrained dipoles (Mosher et al., 1992) at any timepoint, using the three-layered concentric sphere head model by the nonlinear optimization methods (Kamijo et al., 2001). An unconstrained dipole was estimated at any timepoint with maximal peak or trough in the EEGs reconstructed by the deflation procedure for each IC. Here, we searched for appropriate and reliable dipole solutions, by selecting localization results only with goodness of fit (GOF) of more than 90% and with the simplified confidence limits (CLs) (Yamazaki et al.,

2000) of less than 1 mm, by restricting to the results with no drastic change in the brain sites where the unconstrained dipoles are located at least twenty successive instants including the peak or trough, and by excluding the ECDL results localized to the cerebral ventricles and the corpus callosum.

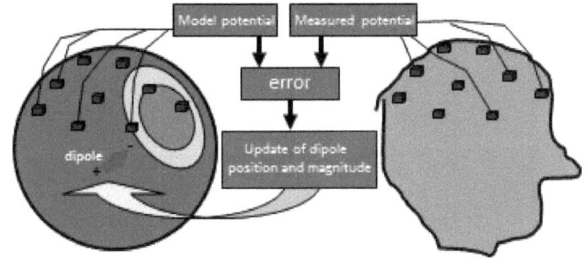

Comparing the human head to the earth, when earthquake occurred,
Seismometer   ⇒electrodes
Seismograph   ⇒EEGs
Magnitude and position of the focus⇒those of dipoles

Fig.2-7: Concept of equivalent current dipole source localization, comparing the human head to the earth, when earthquake occurred.

Anatomical labeling of the brain where ECDs were located, using the Japanese brain atlas for a single subject, was automatically carried out in the following: each subject's MRI was transformed into the atlas, then the estimated ECDs were projected onto the atlas by this non-linear transformation, and finally anatomical labels on the atlas were determined (Tanaka et al., 2010).

2.1.3.3 Spectrograms

Speech signals could be transformed into spectrograms by short-time Fourier transform. Fig.2-8 shows an example for the Fourier transform of one speech signal. This spectral has some maximum values. From early maximum value, the first formant frequency (F1), the second one (F2), the third one (F3), ... are called. Spectrograms refer to the spectral with time course. Here, we use only F1 and F2.

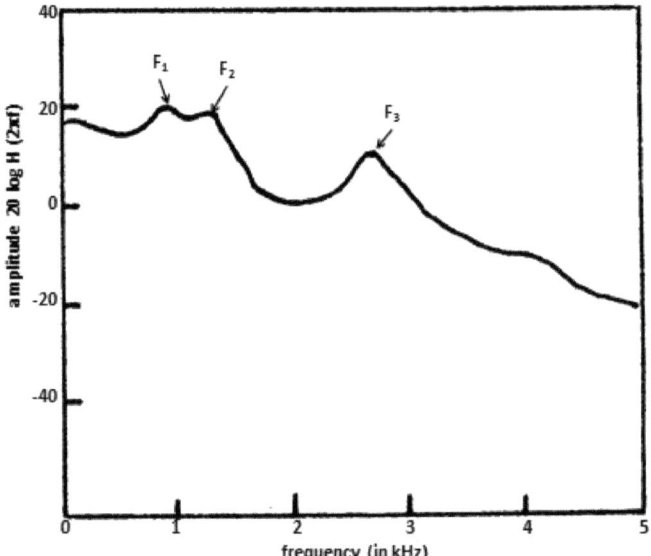

Fig.2-7: Power spectral of one speech signal.

2.1.3.4 Kalman filter

Next, according to the hypothesis, assumed in Directions Into Velocities of Articulators (DIVA) model (Guenther et al., 2006; Guenther et al., 2009), that neurons in the left ventral premotor cortex present intended speech sounds in terms of formant frequency trajectories, and projections from these neurons to the primary motor cortex transform the intended formant trajectories into motor commands to the speech articulators, the relationship between the extracted ICs and spectrograms of the speech signals was described by a Kalman filter. The filter was given by

$$x_t = Ax_{t-1} + w_t$$
$$y_t = Cx_t + v_t,$$

where the parameters A, C, w and v were those to be estimated (Wu et al., 2006), where x_t is the two-dimensional vector consisting of the first (F1) and second (F2) formant frequencies, y_t is the one-dimensional vector representing one IC, the matrix A describes the relationship between past and future formant frequencies, C describes the expectation of the reconstructed EEGs given a set of formant frequencies and the error terms w_t and v_t are white Gaussian random variables.

In the decoding phase (Fig.2-9), the inputs to the Kalman filter specified in the learning phase were the ICs whose dipole solutions were located at the premotor cortex, SMA and/or Broca's area, according to the previous neuroimaging studies related to silent speech (Shergill et al., 2002; Keller et al., 2003; Shuster and Lemieu, 2005; Abe et al., 2011; Van der Haegen et al., 2013), and the filter estimated *spectrograms* for the silent speeches using the so-called Kalman filter algorithm (Kalman, 1960). The 0 ms on the EEGs was defined to be average EMG onset in the learning phase for each subject (Fig.2-9).

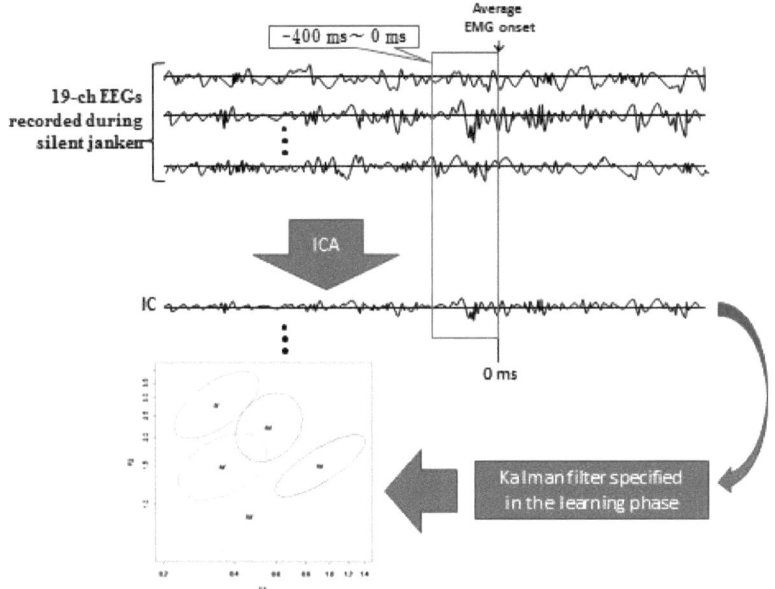

Fig.2-9: Decoding phase in the present SSBCIJ.

2.2 Results
2.2.1 Grand averages

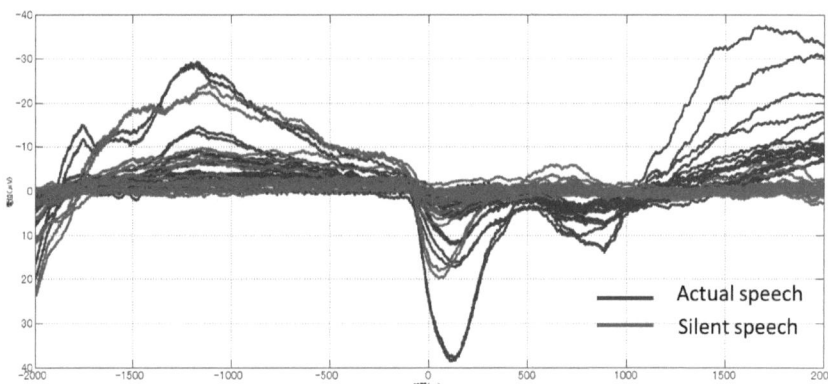

Fig.2-10: Grand averages of the 19-ch EEGs recorded during the actual (blue line) and silent (red one) janken tasks.

Fig.2-10 shows the grand averages for the actual and silent janken tasks. This figure reveals similar Bereitschaftspotential (BP)-like component both for the actual and silent speeches, while motor potential (MP)-like ones (Shibasaki and Hallett, 2006) for only the actual one. Therefore, for both the tasks, we should pay attention to BP-like components (Deecke et al., 1986).

2.2.2 Correlation coefficients

We statistically examined the hypothesis in the DIVA model for our SSRS. From the correlation coefficients, with the minimal P-values, for all the tasks by all the subjects, between each formant frequency and the IC, it followed that significant correlations were found for both F1 ($r=0.62\sim0.88$, $p \leq 6.00\times10^{-15}$ for "rock"; $r=0.64\sim0.94$, $p \leq 7.64\times10^{-13}$ "paper"; $r=0.54\sim0.95$, $p \leq 3.27\times10^{-13}$ for "scissors") and F2 ($r=0.65\sim0.92$, $p \leq 7.55\times10^{-15}$ for "rock"; $r=0.60\sim0.98$, $p \leq 2.83\times10^{-11}$ for "paper"; $r=0.68\sim0.92$, $p \leq 7.88\times10^{-15}$ for "scissors") except for one subject (DK) (Table 2-1), thus confirming the hypothesis. The Kalman filter parameters were calculated by using one pair of the ICs and the spectrogram, whose correlation coefficient had the minimal P-value.

2.2.3 Estimated spectrograms for the silent janken

As the training performance, in case of one subject, diagonal parts of Fig.2-11 show the predicted spectrograms in the F1-F2 plane for the silent "rock" , "paper" and "scissors" with ellipsoidal distributions of five Japanese vowels (Kasuya et al., 1968). In case of "rock" (/gu:/) and "paper" (/pa:/), when the formant frequency trajectories reach the /u/ and /a/ regions, respectively, the predictions were considered to be correct, while, in case of "scissors" (/tʃɔki/), the trajectory was regarded as right if it passed through the region /o (ɔ)/ then the distribution /i/. In terms of Japanese pronunciation, a main difference between "scissors" and the others is that the former has two different vowels, and the latter one. To incorporate this difference in the Kalman filter algorithm, the initial values of the covariance matrix (Kalman, 1960) were set to be variances of F1 and F2 and their covariance. Fig.2-12 plots all the spectrograms for each janken in the F1-F2 plane, including the covariance matrices. Note that the covariance for "scissors" was much larger than those for "rock" and "paper". The diagonal parts of Fig.2-11 shows the outputs from the Kalman filter algorithm with the initial values (V) depicted in Fig.2-12, and all indicates the correct predictions. The same tendencies as in Fig.2-11 were obtained for all the rest subjects. The rest of Fig.2-11 exemplifies the misapplication of our predictors. For example, the "rock" predictor correctly estimated only for the silent "rock" EEGs.

Table 2-1: Correlation coefficients between each formant frequency (F1 and F2) and the IC, where the width of a time window is 300 ms, and the starting time for the calculations. IC1 or IC2 depicts the starting time (in ms) in the IC for the calculation of correlation coefficients (r), with the minimal P-values, between the IC and F1 or F2, where, in this table, 0 ms is defined to be 400 ms before the EMG onset. FF1 or FF2 represents the starting time (in ms) in the F1 or F2 for the calculation.

"rock"							"paper"							"scissors"					
subjects	IC1	FF1	r	IC2	FF2	r	subjects	IC1	FF1	r	IC2	FF2	r	subjects	IC1	FF1	r	IC2	FF2	r
DK	97	172	0.82	290	91	0.77	DK	161	175	0.85	189	143	0.85	DK	245	192	0.78	236	134	-0.68
HT	270	249	0.76	264	156	0.84	HT	75	77	0.81	21	1	0.86	HT	270	257	0.81	251	88	0.88
ST	185	185	0.81	197	15	0.92	ST	297	47	0.91	205	286	0.94	ST	213	98	0.95	229	47	0.92
MS	246	62	0.75	212	207	0.78	MS	258	19	0.94	281	42	0.98	MS	118	20	0.81	204	114	0.79
KY	106	43	0.83	75	50	0.79	KY	98	61	0.88	96	231	0.94	KY	190	292	0.81	116	282	0.79
TS	183	12	0.80	233	205	0.83	TS	50	277	0.86	59	213	0.83	TS	261	34	0.79	251	15	0.82
KM	273	54	0.88	152	205	0.83	KM	208	274	0.88	143	143	0.77	KM	226	300	0.88	122	233	0.74
SS	166	301	0.62	268	301	0.69	SS	140	167	0.64	4	24	0.60	SS	174	259	0.65	35	202	0.66
HI	65	112	0.71	278	100	0.65	HI	30	7	0.79	23	265	0.77	HI	84	192	0.54	84	65	0.56
KT	149	169	0.78	1	165	0.79	KT	158	301	0.88	100	104	0.78	KT	160	11	0.86	132	227	0.70

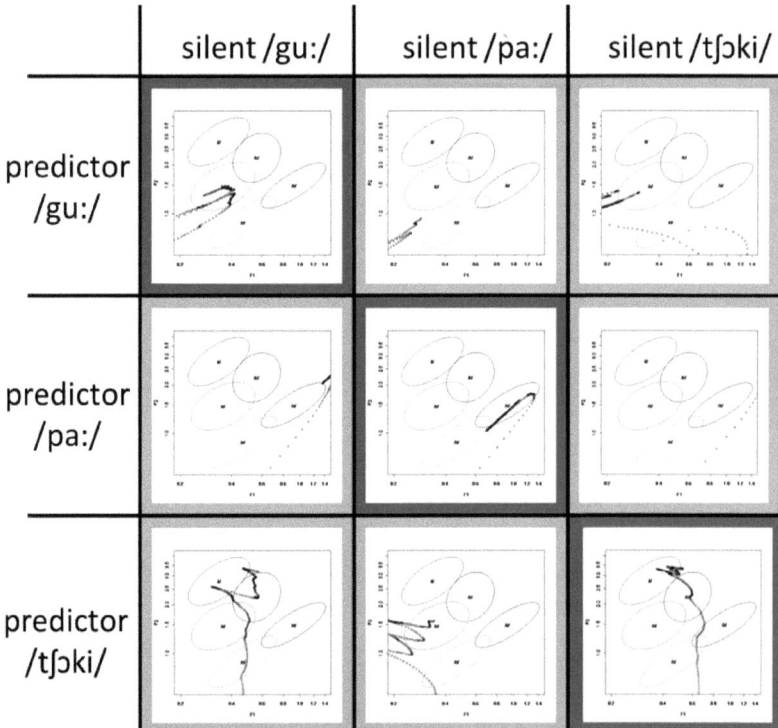

Fig.2-11: A confusion matrix of the three silent janken predictors for one subject.

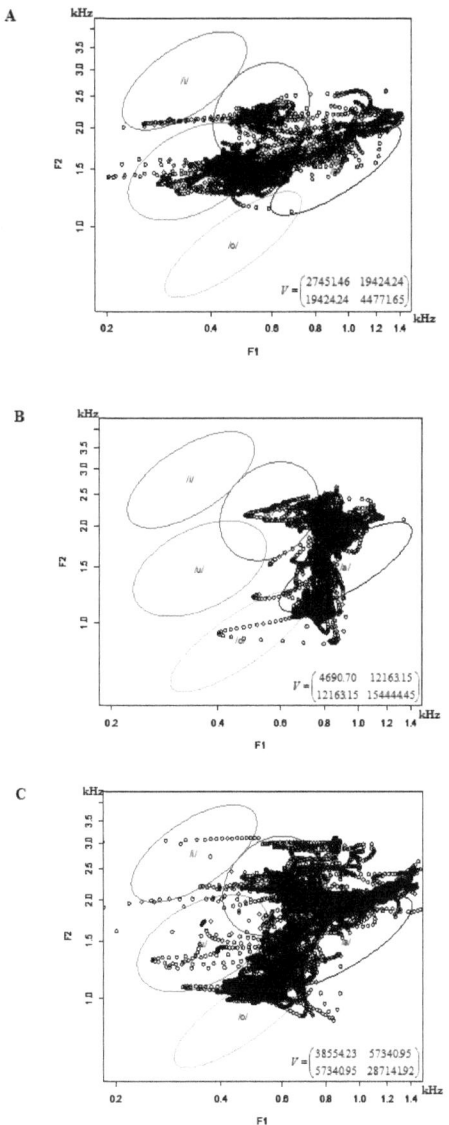

Fig.2-12: Spectrograms in the F1-F2 plane, obtained from the speech signals during the actual janken tasks: (A) "rock"; (B) "paper"; (C)

"scissors". Each task includes 30 trials.

2.3 Discussion

In order to decode silent speeches from single trial EEGs, we used Kalman filters for the vowel recognition. The performance of the present Kalman filters might be improved in the following.

2.3.1 Three-dimensional Kalman filter

By constructing three-dimensional Kalman filter, that is, involving F3, we obtained more discriminative results for the silent "rock" and "paper" tasks (Figs. 2-13 (A) and (B), respectively).

A

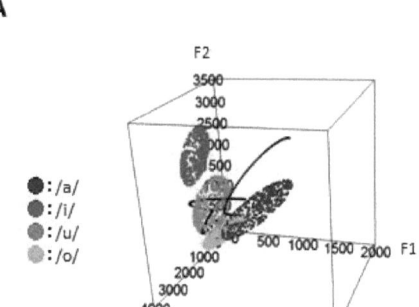

B

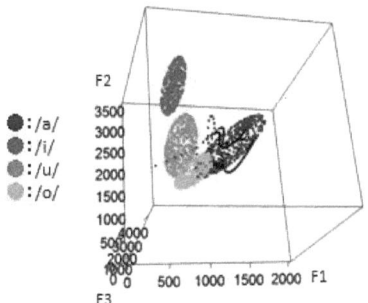

Fig.2-13: Spectrograms (black points) predicted from the three-dimensional Kalman filter for the EEGs recorded during the silent janken tasks: (A) "rock"; (B) "paper".

2.3.2 Particle filter

In KFs, the relationship between the brain activity and speech signals is assumed to be linear in this study. This linearity might limit to the accuracy of the present JSSBCI. For the purpose, the KF and particle filter (PF) were compared in the following. The PF is a technology for implementing a recursive Bayesian filter by Monte Carlo simulations, whose key idea is to represent the probabilistic density function to be required by a set of random samples (particles) with associated weights. The PF based on Monte Carlo simulation has parametric state vectors, and can define likelihood for observation values and apply to non-linearity.

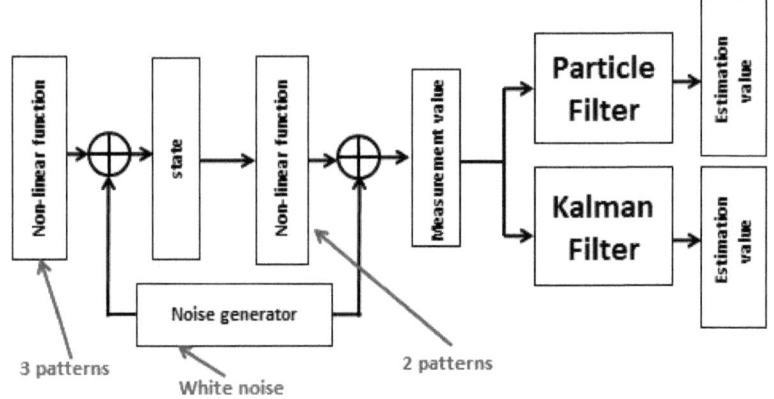

3 patterns

White noise

2 patterns

50 times of estimation for each pattern
Validation by MSE
Using MATLAB

Fig.2.14: Computational simulations for comparing Kalman and particle filters.

Here, we tried computational simulations shown in Fig.2-14 (Shono 2015). By generating state vectors using three kinds of non-linear functions with white noises, and estimating measurement values using two kinds of non-linear functions with white noises, the two filters (KF and PF) were validated by mean square errors (MSEs).

Fig.2-15 shows an example on the validation, demonstrating that the PF was significantly nearer the true states than the KF ($p<0.05$) (Shono 2015). Consequently, such examples were yielded for 5 out of 6 patterns. This simulation result implies that when fitting to optimal non-linear functions and defining optimal likelihood functions, the PF could more accurately estimate than the KF.

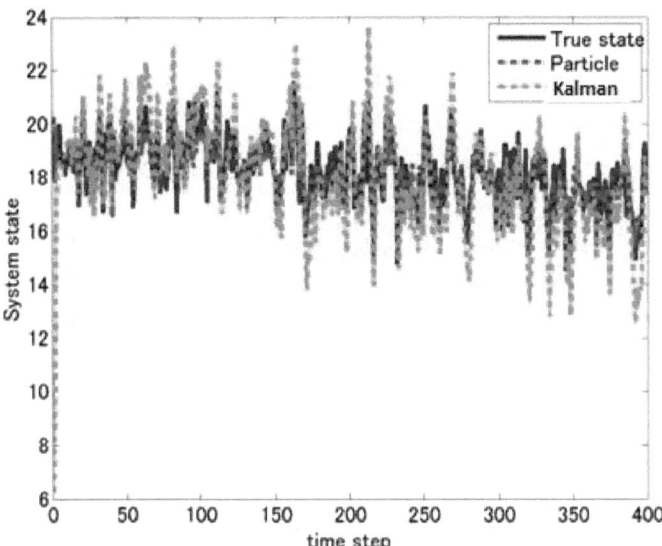

Fig.2-15: An example for the validation of the KF and the PF for the true states.

Chapter 3 Experiment II: Silent Word Recognition

Including Consonants

The SSBCIJ for silent janken in Experiment I was constructed in terms of vowel recognition. Therefore, for example, spring ("haru" in English pronunciation of Japanese) and summer ("natsu" in English one) could not be discriminated by the JSSBCI, because the vowel transitions are the same. In order to cope with this problem, Experiment II was designated. In the learning phase for Experiment II, speech signals were transformed into vowel and consonant sequences, and these transitions were learned by hidden Markov model (HMM). In the decoding phase, the inputs to the HMM are *spectrograms* estimated from the Kalman filter specified in the learning phase. Which season was silently spoken was determined by the maximal likelihood among each HMM output. Moreover, mel-frequency cepstral coefficients (MFCCs) are used as speech signal features, in addition to spectrograms.

3.2 Materials and Methods
3.2.1 Subjects

Six student volunteers (22-28 years; one female) with no history of neurological disorders participated in Experiment II, whose procedures were also approved by the Ethics Committee for Human Subject Research, Faculty of Computer Science and Systems Engineering, Kyushu Institute of Technology. Informed consents were obtained from all the students in writing for the procedures prior to the experiment. All the volunteers were right-handed except for one male and all native Japanese speakers.

3.1.2 Tasks and EEG and speech signal data acquisition

Any of four kinds of photographs was presented on the monitor. These photographs could be associated with "spring", "summer", "autumn" and "winter" by the subjects. The subjects were requested to overtly and covertly speak the season associated in Japanese.

After the subject gazed for 2.5 s a point presented at the center of a monitor 62 cm away from the subject, a landscape photograph being associated with "spring", "summer", "autumn" or "winter" was presented for the next 3 s. Only the fixation point was presented for the next 2.5 s. Then, when the point disappeared, the subject overtly or covertly speaks the corresponding season (Fig.3-1). The photographs were randomly presented ten times for each season.

Thirteen active electrodes (AP-C100-0155, DIGITEX LAB. CO., LTD., Japan) were affixed to the scalp according to the International 10-20 Systems (F3, F5, F4, F6, F7, F8, FC3, FCz, FC4, C3, Cz, C4, POz). Additive two channels were included for electromyography (EMG) monitoring face and mouth movements. The EEGs recorded at each electrode were fed to an amplifier (Polymate AP1132, DIGITEX LAB. CO., LTD., Japan) with 10000 gain and a notch filter of 60 Hz. The amplified EEGs were sampled at a rate of 1 kHz during an epoch of 3 s preceding and 3 s following each stimulus presentation. The online A/D converted EEG data was immediately stored on a hard disk in a personal computer.

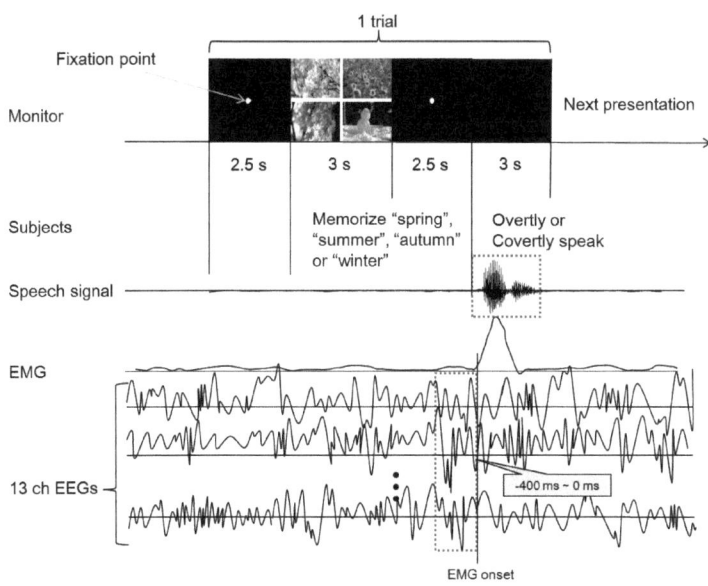

Fig.3-1: Time-scheduling of stimulus presentation, tasks and measurements of speech signal, 13-ch single-trial EEGs and EMGs.

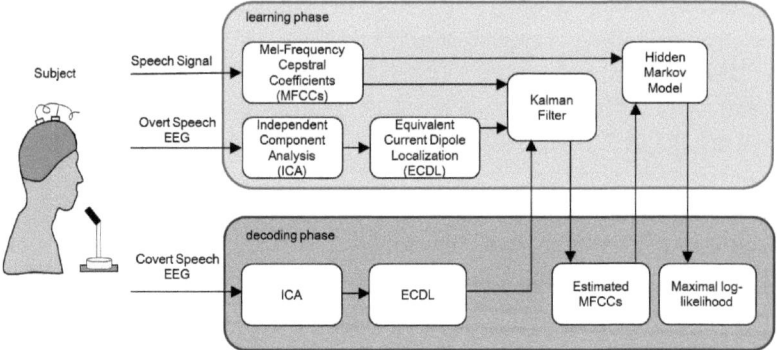

Fig.3-2: An overview of the present SSBCIJ system. The system runs in two phases. In the learning phase, we instructed the subjects to perform overt speech tasks and collected EEG, EMG and speech signal. The Kalman filter and HMM were learned using these data. In the decoding phase, the silent speech EEG was inputted to the Kalman filter and HMM obtained in the learning phase, then which "season" was covertly spoken was determined by the likelihood as each HMM output.

3.1.3 Data processing

As summarized in Fig.3-2, the present SSBCIJ also consists of two phases.

3.1.3.1 ICA and ECDL

In the learning phase, at first, independent component analysis (ICA) (Hyväreinen and Oja, 1997) then equivalent current dipole source localization (ECDL) (Kamijo et al., 2001) were applied to single-trial EEGs recorded during the actual speech task, and ICs were extracted whose dipole solutions were located at the Broca's area. The time intervals in the EEGs to be analyzed were from -400 ms to 0 ms that correspond to negative slope (NS') of Bereitschaftspotential (BP) (Deecke et al., 1986), where 0 ms was defined to be the EMG onset (Fig.3-2). For the onset of silent speech, the average EMG onset of actual speeches was used.

3.1.3.2 Spectrograms and mel frequency cepstral coefficient

Speech signals can be transformed into spectrograms by short-time Fourier transform, and into cepstral. The speech signals can be regarded as the convolution of sound source signals such as turbulence by oscillation and friction of the vocal cords and articulation filter determined by the shapes of the vocal tract and the nasal and oral cavities (Fig.1-1). The cepstral analysis is a method for dividing the speech signals into sound source signals and articulation filter, with reference to that the articulation filter more smoothly changes in frequency domain than the sound source signals. For the power spectral obtained as Fourier transform of the speech signals, the cepstrum refer to the logarithmic transformation then the inversed Fourier transform. The unit of the cepstrum is called quefrency. The characteristics of the articulation filter and the sound source signal are revealed in low quefrency domain and in high one, respectively.

Mel-frequency cepstral coefficient (MFCC) is the weighted cepstral by mel scale based on human frequency perception characteristic. 39-dimensional MFCC contains 12-dimensional one, one-dimensional normalized log-energy, 13-dimensional delta coefficient and 13-dimensional acceleration coefficient. Speech signals were downsampled to 1.6 kHz by Wavesurfer (Wavesurfer 1.8.8p4, http:// www.speech.kth.se /wavesurfer/) and transformed into the MFCCs by Hcopy of Hidden Markov Toolkit (Hidden Markov Model Toolkit (HTK) 3.4, http://htk.eng.cam.ac.uk/).

3.1.3.3 Kalman filter

The relationship between the extracted ICs and the MFCCs was described by Kalman filters (KFs):

$$x_t = Ax_{t-1} + w_t$$
$$y_t = Cx_t + v_t,$$

where the parameters A, C, w and v were those to be estimated (Wu et al., 2006). Where x_t is the 39-dimensional vector consisting of MFCCs, y_t is the one-dimensional vector representing an IC whose dipole sources were located at the Broca's area, the matrix A describes the relationship between past and future MFCCs, C describes the expectation of the reconstructed EEGs given a set of MFCCs and the error terms w_t and v_t are white Gaussian random variables $N(0,Q)$ and $N(0,R)$, respectively, with Q the residual covariance of the state linear dynamical system and the R the residual covariance of the likelihood function.

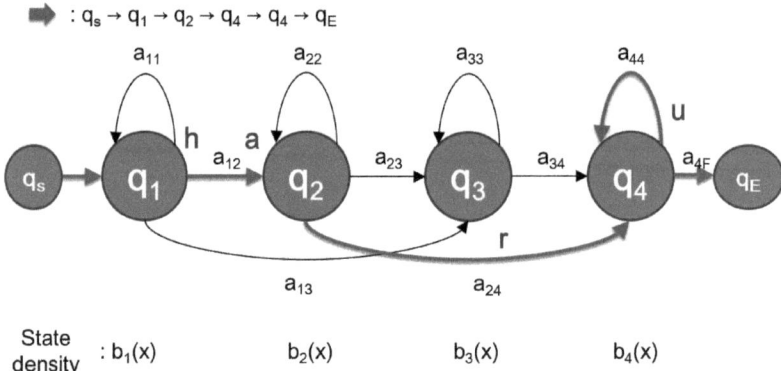

Fig.3-3: Representation of the present left-to-right HMM with six states.

3.1.3.4 HMM
3.1.3.4.1 HMM using spectrograms

For example, the present "spring"-HMM is the left-to-right one shown in Figure 6, and is characterized by the following: (i) N is the number of states in the model; (ii) M is the number of output vowel and consonant; (iii) $A = (a_{ij})$, the transition matrix of the underlying Markov chain, where a_{ij} is the probability of making a transition from state i to state j; (iv) $b_j(x_t)$ is the probability of outputting digitalized speech signal x_t in state j; (v) $\pi = (\pi_i), i = 1, 2, \cdots, N$, the initial state probability vector. Fig.3-3 exemplifies $N=5$ and $M=4$. In the learning phase, each vowel and consonant occurrence is segmented into N states. This segmentation is achieved by finding the optimum state sequence, via the Viterbi algorithm with $b_j(x_t)$ modeled by Gaussian mixture densities (Fig.3-4), in addition to the initialization of π_i and a_{ij}. Parameters in the densities are estimated from spectrograms for ten actual speech trials after K-means clustering (Rabiner et al., 1985). That is, for $i=1, 2,\ldots, N$,

$$f'(i,t) = \begin{cases} \log \pi_i & (t = 0) \\ \max_j \left\{ f'(i,t-1) + \log a_{ji} b_j(y_t) \right\} & (t = 1,2,\ldots,T) \end{cases}$$

are calculated, then a maximum log-likelihood

$$L = \max_{i,\, q_i \in F} f'(i,T)$$

is obtained, where F is a set of final states. Thus, the HMM parameters were initialized by the Viterbi algorithm and then re-estimated by the Baum-Welch algorithm. These procedures were carried out by HTK (a portable toolkit for building and manipulating HMMs in C: http://htk.eng.cam.ac.uk/). In the decoding phase, a silently spoken season was assumed to be maximal among each season-HMM likelihood value for the predicted spectrogram from the Kalman filter.

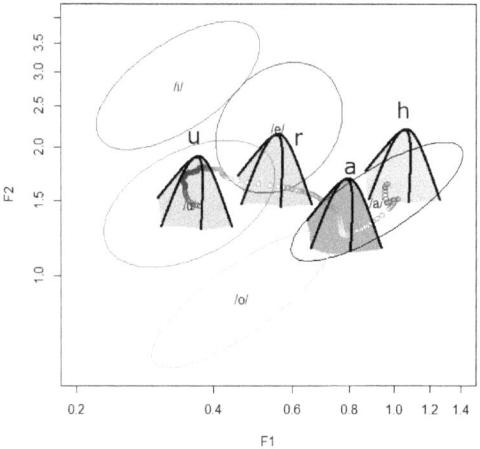

Fig.3-4: A Gaussian mixture density model for "haru" on the F1-F2 plane.

3.1.3.4.2 HMM using MFCCs

Moreover, using the MFCCs, a left-to-right hidden Markov model (HMM) with six states, for example, in case of "spring" (/haru/), the HMM was constructed by four phonemes and beginning and ending states (Fig.3-5).

The decoding phase carried out the same procedure as the learning phase for single-trial EEGs measured during the silent season tasks. The extracted ICs were inputted to the Kalman filter obtained in the learning phase. Then, the Kalman filter estimated silent speech MFCCs according to the so-called Kalman filter algorithm. Finally, the likelihood for each silent season trial was calculated by the HMM, the input to whom was the estimated MFCC. Which "season" was silently spoken was determined by the maximal likelihood.

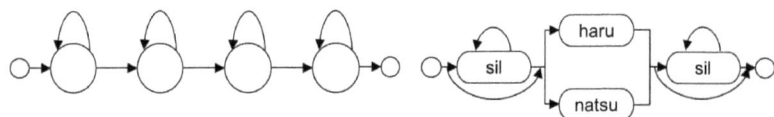

Fig.3-5: *left:* HMM design. Six-states including 2 dummy states at the beginning and the end, and left-to-right model without skip over states. *right:* HMM network. The network means that there are 0 or more times soundless intervals before and after the speech interval.

3.2 Results

The following results were reported only on a few right-handed subjects (22-23 years) for the silent "haru" and "natsu", which are the same vowel transitions, recognition and discrimination.

3.2.1 HMM using spectrograms

Table 3-1 shows a confusion matrix of haru- and natsu-HMMs using spectrograms in terms of log-likelihood values for silent haru and natsu trials by one subject, which demonstrates that our SSBCIJ correctly recognized the trials. The accuracy was 39 % and 83 % for 10 trials during the silent haru tasks and those during the silent natsu ones, respectively.

Table 3-1: A confusion matrix of haru- and natsu-HMMs.

	haru-HMM	natsu-HMM
haru-SSEEG	-3033.3	-3136.1
natsu-SSEEG	-5018.0	-4718.4

Table 3-2 indicates a confusion matrix in case of that which task was executed is unknown, which means that even this situation could be discriminated by the present HHMs using spectrograms.

Table 3-2 A confusion matrix in case of that which task was executed is unknown.

	haru-Kalman filter		natsu-Kalman filter	
	haru-HMM	natsu-HMM	haru-HMM	natsu-HMM
haru-SSEEG	-3033.3	-3136.1	-3196.4	-3160.6
natsu-SSEEG	-5819.5	-4916.0	-5018.0	-4718.4

Table 3-3 shows a confusion matrix for all the silent "season" tasks (Itoh et al., 2015). For example, at the first row (Silent "spring" EEG), for the estimated silent "spring" spectrograms, the log-likelihood of the "spring"-HMM was higher than those of the other-HMMs. The other rows also demonstrate the same tendency. Therefore, if higher log-likelihood values are accepted, it could be demonstrated that these HMMs work well. As a preliminary result, the accuracy was 86% ("spring"), 29% ("summer"), 43% ("autumn") and 100% ("winter") for one subject.

Table 3-3: A confusion matrix for all the silent "season"-HMMs.

	Silent "spring" EEG	Silent "summer" EEG	Silent "autumn" EEG	Silent "winter" EEG
"spring" – HMM	**-2844**	-5749	-4663	-6441
"summer"- HMM	-2878	**-5088**	-4246	-6064
"autumn"- HMM	-3377	-5429	**-4204**	-6055
"winter" - HMM	-3310	-5940	-5609	**-5510**

3.2.2 HMM using MFCCs

MFCC estimated for each of two silent season trials was inputted to the haru- and natsu-HMMs both obtained in the learning phase. Table 3-4 shows the outputs (log-likelihood) from the HMMs. Because of -3390.69 > -6999.30 and -4021.47 > -29395.00, this table demonstrates that each HMM correctly discriminated two kinds of silent season trials. Such correct confusion matrices were obtained for 5 out of 11 trials and 8 out of 9 ones in the silent haru and natsu EEGs, respectively. That is, the accuracy was 0.45 and 0.89, respectively (Yamaguchi et al., 2015b).

Table 3-4: A confusion matrix of the HMMs using MFCCs.

	silent speech /haru/	silent speech /natsu/
haru-HMM	-3390.69	-29395.00
natsu-HMM	-6999.30	-4021.47

3.3 Discussion

The present reports were not so good in terms of accuracy. So, we are trying the improvement in the following two.

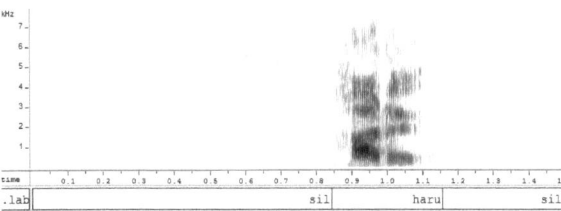

Fig.3-6: Speech signal /haru/ including soundless interval before and after the utterance.

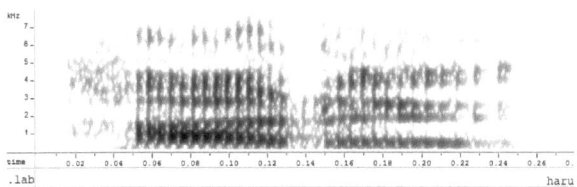

Fig.3-7: Speech signal /haru/ obtained by removing the soundless intervals in Fig.3-6.

3.3.1 Removal of soundless interval from speech signals

The Kalman filter parameter estimation requires the same number of sampling points for the ICs and the MFCCs. Consequently, speech signals to be analyzed were forced to contain soundless intervals, because the MFCCs were 39-dimensional. Moreover, the likelihood for the soundless intervals was generally known to be low. So, in the learning phase, the first 13-dimensional MFCCs were calculated from the speech signals where the soundless intervals were removed from the speech signals, and then the Kalman filter parameters were calculated from the MFCCs and the ICs whose dipole source was located at the Broca's area. In the decoding phase, the Kalman filter estimated the 13-dimensional MFCCs using the EEGs recorded during the silent season tasks. Next, the rest 26-dimensional MFCCs were predicted from the estimated 13-dimensional ones by the regression analysis (Furui, 1986). Finally, each HMM outputted log-likelihood using all the MFCCs. The same results as in Table 1 were obtained in 10 out of 11 trials and 5 out of 9 ones in the silent haru and natsu tasks, that is, the accuracy was 0.91 and 0.56, respectively (Yamaguchi et al., 2015b).

3.3.2 Averaged Kalman filter parameters

For even any one trial in the learning phase, if we would obtain ICs whose dipole solutions were localized to the Broca's area, the corresponding Kalman filter could be applied to any one trial with silent season EEGs having the Broca's area dipoles. Different silent-season trial EEGs were applied to different Kalman filter. Therefore, the difference in Kalman filters among the trials in the learning phase might not yield the good accuracy. So, the Kalman filter with averaged parameters on all the single-actual-speech-trial EEGs, whose dipole sources were localized to the Broca's area, in the learning phase was applied to the silent-season-trial EEGs with the Broca's area dipoles.   With the soundless interval removal from the speech signals, the averaged Kalman filter correctly estimated 10 out of 11 trials and 8 out of 9 ones in the silent haru and natsu, that is, the accuracy was 0.91 and 0.89, respectively (Yamaguchi et al., 2015b).

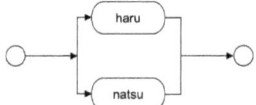

Fig.3-8: Improved HMM network without sil-HMM by removing the soundless intervals.

Chapter 4 Further Researches

In the present SSBCIJ, there are many parameters to be optimized as follows:

(1) Characterization of speech signals
(2) Number of ICs
(3) EEG intervals to be analyzed by ECDL for KF construction
(4) Is it sufficient only to use the BA-ICs?
(5) What does the brain activity encode in speech?
and so on.

4.1 Characterization of speech signals

In this study, speech signals were transformed into formant frequencies and MFCCs. The former is better in terms of efficacy, the latter better in terms of accuracy. However, we will pay attention to MFCCs in future, because we should make much account of the accuracy.

4.2 Number of ICs

Intrinsically, vowels and consonants are known to be processed by distinct neural mechanisms (Caramazza et al., 2000). For example, vowels and consonants increased activation in right middle temporal and frontal areas, respectively (Carreiras and Price, 2008). Tentatively, we constructed a Kalman filter with one IC whose dipole solution was located at the temporal area, in addition to the frontal-area-dopole IC, in the learning phase, and then the silent "haru" spectrogram was estimated in the learning phase. Figs.4-1 (A), (B) and (C) show spectrograms obtained by the Kalman filter with only one IC, that with the above two ICs and that with two ICs whose dipoles were localized to the other areas, respectively. Fig.4-1 (B) revealed the best performance.

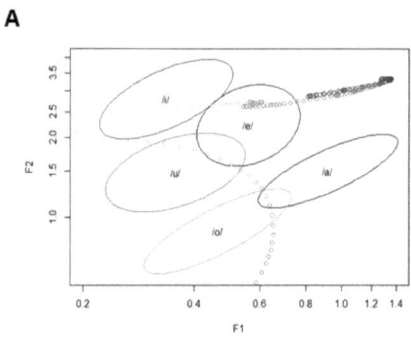

A

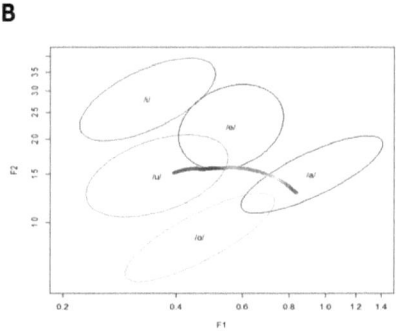

B

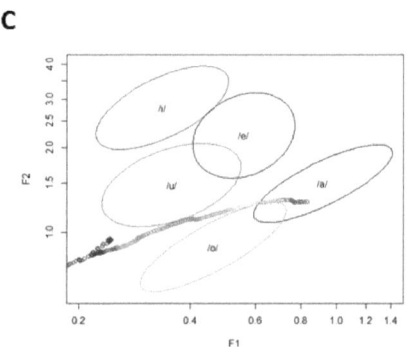

C

**Fig.4-1: Spectrograms obtained from the Kalman filter with only
one IC (A), that with two ICs corresponding to the Broca's and**

Wernicke's areas (B) and that with two ICs whose dipoles were located at the other areas (C).

4.3 One-hiragana-character-HMM (1-HC-HMM)

A   B

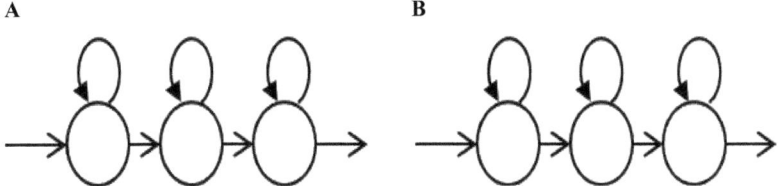

Fig.4-2: ha- (A) and ru- (B) HMMs.

In future, we must step up to silent word recognition, in addition to solving practical problem. For the purpose, we are now engaged in construction one-hiragana-character-HHMs (1-HC-HMM) (Hirose et al., 2015). These HMMs have two phoneme states (Fig.4-2). Moreover, in the learning phase, KFs were constructed using not subjects' speech signal but speech signal database (e.g., Speech Resources Consortium (http://research.nii.ac.jp/src/en/). Table 4-1 (A) shows a preliminary result on 1-HC-HMM. The accuracy was not good.

Table 4-1: 1-HC-HMM accuracy: (A) no limitation to any frequency band; (B) limitation to gamma band.

A

1-HC-HMM	accuracy
ha	0/10
ru	6/10
na	7/8
tsu	1/8

B

1-HC-HMM	accuracy
ha	9/10
ru	5/10
na	8/8
tsu	0/8

4.4 EEG intervals to be analyzed by ECDL for KF construction
4.4.1 Determination of EMG onset

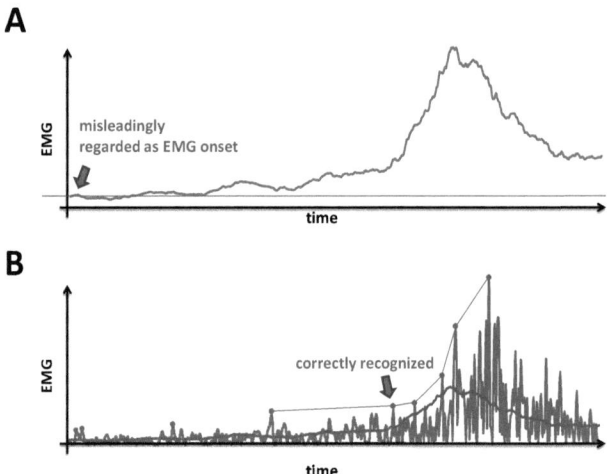

Fig.4-3: Two methods for determining EMG onset: (A) the previous one; (B) the proposed one.

In this book, the EEG interval to be analyzed by ECDL had been limited to the NS' of the BP (Shibasaki and Hallett, 2006). The NS' refers to the interval between -400 ms and 0 ms, where the 0 ms is defined to be EMG onset. Generally, the EMG onset is regarded as the time when 5 % of the maximal EMG amplitude firstly occurred. However, there are some cases where this criterion was not appropriate as exemplifies in Fig.4-3 (A).

Table 4-2: (A) a confusion matrix using both the NS' and the early BP on the basis of improved EMG onset; (B) a confusion matrix for speech signals using HMMs.

A

	HMM_haru	HMM_natsu	HMM_aki	HMM_huyu
EEG_haru	-4106.8	-4800.5	-3498.4	-13334.6
EEG_natsu	-	-	-15071.6	-
EEG_aki	-3983.0	-4423.4	-3866.1	-14483.5
EEG_huyu	-4936.4	-6059.4	-4272.3	-4202.6

B

		HMM_haru	HMM_natsu	HMM_aki	HMM_huyu
	haru	-2913.4	-3517.5	-3361.8	-14329.4
speech signals	natsu	-5651.0	-4790.8	-5129.3	-54206.3
	aki	-3917.6	-3882.7	-3684.6	-21091.3
	fuyu	-6107.7	-7830.3	-4079.1	-3403.3

So, we designated a new criterion for EMG onset in the following. Our idea is to extract maximum values of EMG amplitudes in order and to calculate a slope between two neighboring timepoints having the values. Then, when the slope became beyond a threshold, the EMG onset is defined to occur (Fig.4-3 (B)). The accuracy, where EMG onsets were correctly detected, was 58 % in 100 trials (Matsushita et al., 2015). Moreover, the confusion matrix based on this improvement is shown in Table 4-2 (A), which was better than using only the NS'. Table 4-2 (B) demonstrates that Table 4-2 (A) is similar to Table 4-2 (B) in case of using both the NS' and the early BP (Matsushita et al., 2015).

4.4.2 Inclusion of till early BP

Also based on the results in Experiment I in terms of correlation coefficients, the EEG interval to apply ECDL to has been limited to one of till 400 ms before EMG onset in this book. However, in this case, there were few BA (Broca's area)-ICs. Because the BA is essential for both actual and silent speeches, we might execute ECDL for EEG intervals of till about 2 s before EMG onset, as shown also in Table 4-2 at the previous subsection 4-3-1. Moreover, we should pay attention to gamma band of EEGs. Evidences for this approach are in the following. Recently, electrocorticograms (ECoGs) have systematically investigated at the cortices related to speech production (e.g., Towle et al., 2008; Sahin et al., 2009; Flinker et al., 2015; Herff et al., 2015). Especially, Towle et al. (2008), Flinker et al. (2015) and Herff et al. (2015) studied the gamma activity (70-100 Hz, 70-150 Hz and 70-170 Hz, respectively) of the ECoGs. Table 4-1 (B) shows the accuracy for 1-HC-HMM, which revealed drastic improvement compared with Table 4-1 (A).

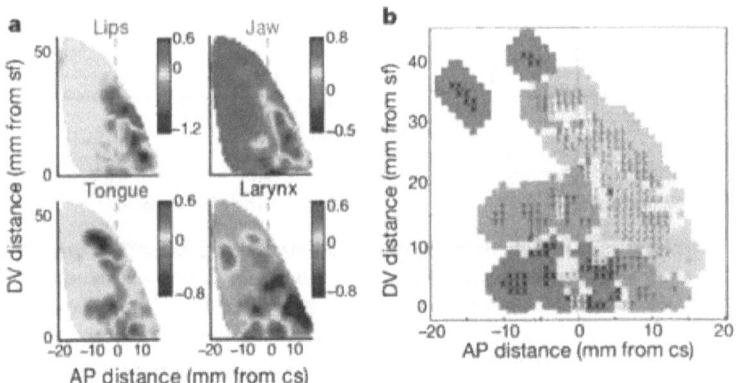

Fig.4-4: Spatial representation of articulators. A, Localization of lips, jaw, tongue and larynx representations. Average magnitude of articulator weightings (colour scale) plotted as a function of anteroposterior (AP) distance from the central sulcus and dorsoventral (DV) distance from the Sylvian fissure (n=3 subjects). B, Functional somoatotopic organization of speech-articulator representations invSMC. Lips (L, red); jaw (J, green); tongue (T, blue); larynx (X, black); mixed (yellow). Letters correspond to locations, based on direct

4.5 Is it OK to construct KFs using only ICs whose dipole solutions were localized to the Broca's area?

In this book, we have consequently utilized the BA-ICs. Is this strategy correct or not? Recently, the Broca's area, which Broca had discovered, have been re-checked (Dronkers et al., 2007). According to this report, the so-called BA is functionally wider than one based on the previous findings, because the brain site involves the insula (IN). So, in future, we should take the IN-ICs into account.

4.6 What does the brain activity encode in speech?

In the last 20 years, there has been an explosion of research into the neural basis of language processing with PET, fMRI, MEG and EEG. Especially, covert articulation activates mainly the Broca's area and the premotor cortex, and other regions, depending on the silent speech tasks, involving the insula, the supplementary motor area (SMA), the pre-SMA-proper, the planum temporale and the inferior parietal lobe (Price 2012).

For example, using high-resolution, multi-electrode cortical recordings, Bouchard et al. (2013) determined the organization of speech sensorimotor cortex in humans during the production of consonant-vowel syllables. They found speech-articulator representations that are arranged somatotopically on ventral pre- and post-central gyri, and that partially overlap at individual electrodes (Fig.4-4). These representations were temporally coordinated as sequences during syllable production. Spatial patterns transitioned between distinct representations for different consonants and vowels over tens of milliseconds. However, these findings would not be directly utilized for silent speech.

On the other hand, Sahin et al. (2009) revealed, by neighboring probes within the Broca's area, distinct neural activity for lexical (~200 ms), grammatical (~320 ms) and phonological (~450 ms) processing, identically for nouns and verbs (Fig.4-5). My concern is the temporal relationship between the Broca's area activity till 450 ms and the BP. The relationship will be essential for the discussion in the above subsection 4.3.

4.7 Where should we go for further researches?

Although the 1-HC-HMM will be helpful, there might be "combinatorial explosion" of the 1-HC-HMM in silent word or phrase recognition. In order to solve this problem, we will require efficient representations of EEGs and a novel statistical learning theory. For the former, the results by Sahin et al. (2009), that is, ECoG tri-phasic pattern, reflecting lexical, grammatical and phonological processing in speech, will be instructive.

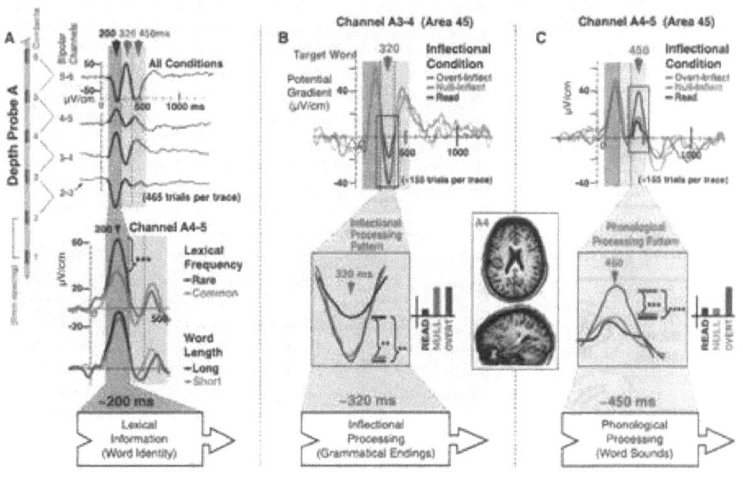

Fig.4-5: (A) Sequential processing of lexical, grammatical, and phonological information in overlapping circuits. (Top) Neural activity recorded from several channels in Broca's area (patient A, Brodmann area 45) shows three LFP components that were consistently evoked by the task (~200, ~320, and ~450 ms). (Bottom) The ~200-ms component is sensitive to word frequency but not word length, suggesting that it indexes a cognitive process such as lexical identification, not simply perception. Stacked waveforms (top and bottom) adopt the axes noted

on the first waveform. (B) At ~320 ms, the LFP pattern suggests inflectional processing. (C) At ~450 ms, in a channel 5 mm distant, the complementary pattern suggests phonological processing. (Inset) MRI slices from this patient, annotated with the anatomical location of A4, the contact in common to the two channels reported here. Statistical significance: **** (P<.0001), *** (P<.001), ** (P<.01) (t test, one tail, two-sample, equal variance). Box arrows (bottom) indicate linguistic processing stages, which may be interposed among other stages not addressed here. Reproduced from Sahin et al. (2009) with permission.

Chapter 5 Conclusions

We started to develop a new single-trial-EEG-based BCI using silent speech in Japanese (SSBCIJ), differently from the previous ones using motor imagery (MI-BCIs). Our JSSBCI is much easier and more natural for users than the MI-BCIs. The key strategy is to construct a model (KF) describing the relationship between speech signals and EEGs recorded during actual (learning phase) and silent (decoding one) speeches, using ICA and ECDL. The EEGs to be analyzed were determined by Bereitschaftspotential (BP) and the ICs whose ECDL solutions were located mainly at the Broca's area. The silent vowel recognition correctly worked in ten healthy, right-handed student volunteers (Chapter 2). The silent word recognition including consonants worked well only in the silent season tasks, but not so good and in few subjects (Chapter 3). Chapter4 addressed itself to the discussions for solving the problems revealed in Chapter 3, and yielded a little improvement. In particular, 1-hiragana-character-HMM (1-HC-HMM) and the BA-ICs limited to gamma band were promising. Nevertheless, "combinatorial explosion" of 1-HC-HMM remains unsolved, and what the BA activity encodes in speech has not been perfectly clarified. I expect young researchers to discover such findings in future. Finally, since Japanese has syllable-timed rhythm (Abercrombie, 1967), it is easy to divide speech signals into phonemes and syllables. My dream to recognize all the combinations of silent hiragana might come true.

References

Abe K, Takahashi T, Takikawa Y, Arai H, Kitazawa S 2011 Applying independent component analysis to detect silent speech in magnetic resonance imaging signals. Euro J Neurosci 34: 1189-1199.

Abercrombie D (Ed) (1967) Elements of General Phonetics. Edinburgh: Edinburgh University Press.

Blakely T, Miller KJ, Rao RP, Holmes MD, Ojemann JG (2008) Localization and classification of phonemes using high spatuial resolution electrocorticography (ECoG) grids. Conference Proceedings of IEEE Eng Med Biol Soc, 4964-4967.

Bos JC, Tack DW (2005) Speech input hardware investigation for future dismounted soldier computer systems. DRCD Toronto CR 2005-64.

Bouchard KE, Mesgarani N, Johnson K, Chang EF (2013) Functional organization of human sensorimotor cortex for speech articulation. Nature 495: 327-332.

Brumberg JS, Nieto-Castanon A, Kennedy PR, Guenther FH (2010) Brain-computer interfaces for speech communication. Speech Commun 52: 367-379.

Callan DE, Callan AM, Honda K, Masaki S (2000) Single-sweep EEG analysis of neural processes underlying perception and production of vowels. Cogn Brain Res 10: 173-176.

Caramazza A, Chialant D, Capasso R, Micell G (2000) Separable processing of consonants and vowels. Nature 403: 428-430.

Carreiras M, Price CJ (2008) Brain activation for consonants and vowels. Cerebral Cortex 18: 1727-1735.

Croft LJ, Rankin PM, Liégeois F, Banks T, Cross JH, Vargha-Khadem F, Baldeweget T (2013) To speak, or not to speak? The feasibility of imaging overt speech in children with epilepsy. Epilepsy Res 107: 195-199.

DaSalla CS, Kambara H, Sato M, Koike Y (2009) Single-trial classification of vowel speech imagery using common spatial patterns. Neural Networks 22: 1334-1339.

Deecke L, Engel M, Lang W, Kornhuber HH (1986) Bereitschaftpotential preceding speech after holding breath. Exp Brain Res 65: 219-223.

Dekins T, Patsis Y, Verhelst W, Beaugendre F, Chapman F (2008) A multi-sensor speech database with applications towards robust speech processing in hostile environments. In: Proc 6th Internat Language Resources and Evaluation (LREC'08), European Language Resources Association (ELRA), Marrakech, Morocco, 28-30 May 2008.

Delorme A, Palmer J, Onton J, Oostenveld R, Makeig S (2012) Independent EEG sources are dipolar. PLoS ONE, 7, issue 2: e30135.

Denby B, Schultz T, Honda K, Hueber T, Gilbert JM, Brumberg JS (2010) Silent speech interfaces. Speech Commun 52: 270-286.

Dronkers NF, Plaisant O, Iba-Zizen MT, Cabanis EA (2007) Paul Broca historic cases: High resolution MR imaging of the brains of Leborgne and Lelong. Brain 130: 1432-1441.

Fagan MJ, Ell SR, Gilbert JM, Sarrazin E, Chapman PM (2008) Development of a (silent) speech recognition system for patients following laryngectomy. Med Eng Phys 20: 419-425.

Falk TH, Paton KM, Chau T (2013) Client-centered music imagery classification based on hidden Markov models of baseline prefrontal hemodynamic responses. In: Brain-Computer Interface Systems – Recent Progress and Future Prospects, Fazel-Rezai R (Ed), Pp.137-154, InTech, Croatia.

Flinker A, Korzeniewska A, Shestyuk AY, Franaszczuk PJ, Dronkers NF, Knight RT, Crone NE (2015) Redefining the rule of Broca's area in speech. PNAS 112: 2871-2875.

Furui S (1986) Speaker-independent isolated word recognition based on emphasized spectral dynamics. Acoustics, Speech, and Signal Processing, IEEE International Conference on ICASSP '86 11: 1991-1994.

Guenther H, Ghosh SS, Tourville JA (2006) Neural modeling and imaging of the cortical interactions underlying syllable production. Brain Lang 96: 280-301.

Guenther FH, Brumberg JS, Wright EJ, Nieto-Castanon A, Tourville JA, Panko M, Law R, Siebert SA, Bartels JL, Andreasen DS, Ehirim P, Mao H, Kennedy PR (2009) A wireless brain-machine interface for real-time speech synthesis. PLoS ONE 4: e8218.

Helmholtz (1853) Ueber einige Gesetze der Verteilung elektrischer Ströme in körperlichen Leitern mit Anwendung auf die thierisch-elektrischen Versuche. Pogg Ann Phsik Chemie 89: 211-233, 353-377.

Herff C, Heger D, de Pesters A, Telaar D, Brunner P, Schalk G, Schultz T (2015) Brain-to-text: decoding spoken phrases from phone representations in the brain. Frontiers in Neuroscience 9: article 217.

Hirose S, Yamaguchi H, Itoh T, Yamazaki T, Fukuzumi S, Yamanoi T (2015) Silent speech BCI – An investigation for practical problems -. IEICE HCG Symposium 2015, 16-18 Dec., Toyama.

Hochberg LR, Serruya MD, Friehs GM, Mukand JA, Saleh M, Caplan AH, Branner A, Chen D, Penn RD, Donoghue JP (2006) Neuronal ensemble control of prosthetic devices by a human with teraplegia. Nature 442: 164-171.

Hueber T, Benaroya EL, Chollet G, Denby B, Dreyfus G, Stone M (2010) Development of a silent speech interface driven by ultrasound and optical images of the tongue and lips. Speech Communication 52:288-300.

Hyvärinen A, Oja E (1997) A fast fixed-point algorithm for independent component analysis. Neural Comput 9: 1483-1492.

Itoh T, Yamaguchi H, Yamaguchi A, Yamazaki T, Fukuzumi S, Yamanoi T (2015) Silent speech recognition system using single-trial EEGs: A silent season BCI. NEUROSCIENCE 2015.

Jung T-P, Makeig S, McKeown MJ, Bell AJ, Lee T-W, Sejnowski TJ (2001) Imaging brain dynamics using independent component analysis. Pro of the IEEE 89 (7): 1107-1122.

Kalman RE (1960) A new approach to linear filtering and prediction problems. J Basic Eng 82: 35-45.

Kamijo K, Kiyuna T, Takaki Y, Kenmochi A, Tanigawa T, Yamazaki T (2001) Integrated approach of an artificial neural network and numerical analysis to multiple equivalent dipole source localization. Frontier Med Biol Eng 10: 285-301.

Kasuya H, Suzuki H, Kido K (1968) Changes in pitch and first three formant frequencies of five Japanese vowels with age and sex of speakers. Acoustical Society of Japan 24: 355-364 (in Japanese).

Keller TA, Carpenter PA, Just MA (2003) Brain imaging of tongue-twister sentence comprehension: Twisting the tongue and the brain. Brain Lang 84: 189-203.

Kellis K, Miller K, Thomson K, Brown R, House P, Greger B (2010) Decoding spoken words using local field potentials recorded from the cortical surface. J Neural Eng 7: 056007.

Kennedy PR, Bakay RAE, Moore MM, Adams K, Goldwaithe J (2000) Direct control of a computer from the human central nervous system. IEEE Trans Rehabil Eng 8(2): 198-202.

Kielar A, Milman L, Bonakdarpour B, Thompson C K (2011) Neural correlates of covert and overt production of tense and agreement morphology: Evidence from fMRI. J Neurolinguistics 24: 183-201.

Kubozono H (Ed) (2010) Phonetics·Phonology. Kurosio Publishers (in Japanese).

Matsumoto M, Hori J (2013) Classification of silent speech using adaptive collection. 2013 IEEE Symposium on Computational Intelligence in Rehabilitation and Assistive Technology (CIRAT) 5-12.

Matsushita K, Hirose S, Itoh T, Nishida H, Yamaguchi H, Yamazaki T (2015) Silent speech BCI − Investigation of algorithm parameters -. Proceedings of Life Engineering Symposium 2015, LE2015, 3A1-3: 41.

Mosher JC, Lewis PS, Leahy RM (1992) Multiple dipole modeling and localization from spatio-temporal MEG data. IEEE Trans Biomed Eng 39: 541-557.

Naci L, Cusack R, Jia VE, Owen AM (2013) The brain's silent messenger: Using selective attention to decode human thought for brain-based communication. J Neurosci 33: 2013-2033.

Nakajima Y, Kashioka H, Shikano K, Campbell N (2003) Non-audible murmur recognition. In: Proc Eurospeech 2003, pp.2601-2604.

Neuper C, Müller GR, Kübler A, Birbaumer N, Pfurtscheller G (2003) Clinical application of an EEG-based brain computer interface: a case study in a patient with severe motor impairment. Clin Neurophysiol 114: 399-409.

Niedermeyer E, Lopes da Silva F (Eds) (1987) Electroenphalography Basic Principles, Clinical Applications and Related Fields. 2nd Edition, Urban & Schwarzenberg·Baltimore-Munich.

Oldfield RC (1971) The assessment and analysis of handedness: The Edinburgh inventory. Neuropsychologia 9: 97-113.

Peeva MG, Guenther FH, Tourville JA, Nieto-Castanon A, Anton J-L, Nazarian B, Alarion F-X (2010) Distinct representations of phonemes, syllables, and supra-syllabic sequences in the speech production network. NeuroImage 50: 626-638.

Price CJ (2012) A review and synthesis of the first 20 years of PET and fMRI studies of heard speech, spoken language and reading. NeuroImage 62: 816-847.

Rabiner LR, Juang B-H, Levinson SE, Sondhi MM (1985) Recognition of isolated digits using hidden Markov models with continuous mixture densities. AT&T Technical Journal 64: 1211-1234.

Riaz A, Akhtar S, Iftikhar S, Khan AA, Salman A (2013) Inter-comparison of classification techniques for vowel speech imagery using EEG sensors. Mathematics and Computers in Science and Industry 31: 309-313.

Sahin NT, Pinker S, Cash SS, Schomer D, Halgren E (2009) Sequential processing of lexical, grammatical, and phonological information within Broca's area. Science 326: 445-449.

Sakamoto M, Yamaguchi H, Yamazaki T, Kamijo K, Yamanoi T (2015) Performance of a Bayesian-network-model-based BCI using single-trial EEGs. IEICE TRANS. INF. & SYST., E98-D(11): 1976-1981.

Shergill SS, Brammer MJ, Fukuda R, Bullmore E, Amaro Jr E., Murray RM, McGuire PK (2002) Modulation of activity in temporal cortex during generation of inner speech. Hum Brain Mapp 16: 219-227.

Shibasaki H, Hallett M (2006) What is the Bereitschaftspotential? Clin Neurophysiol 117: 2341-2356.

Shono T (2015) Evaluation of the estimator used in Silent Speech BCI. Report on the practices during the internship in the Department of Bioscience and Bioinformatics, Kyushu Institute of Technology, 1-2 (in Japanese).

Shuster LI, Lemieu SK (2005) An fMRI investigation of covertly and overtly produced mono- and multisyllabic words. Brain Lang 93: 20-31.

Suppes P, Lu Z-L, Han B (1997) Brain wave recognition of words. PNAS 94: 14965-14969.

Tanaka K, Motoi M, Sasaguri Y, Yamazaki T, Takayanagi H, Yamanoi T, Kamijo K (2010) A new single-trial-EEG-based BCI – Validation of quantification methods of type II modeling. Clin Neurophysiol 121: S161.

Tardelli JD (Ed) (2003) MIT Lincoln Labs Report ESC-TR-2004-084. Pilot Corpus for Multisensor Speech Processing.

Towle VL, Yoon H.-A, Castelle M, Edgar JC, Biassou NM, Frim DM, Spire J.-P, Kohrman MH (2008) EcoG gamma activity during a language task: differentiating expressive and recptive speech areas. Brain 131: 2013-2017.

Van der Haegen L, Westerhausen R, Hugdahl K, Brysbaert M (2013) Speech dominance is a better predictor of functional brain asymmetry than handedness: A combined fMRI word generation and behavioral dichotic listening study. Neuropsychologia 51: 91-97.

Vigneau M, Beaucousin V, Hervé PY, Duffau H, Crivello F, Houdé O, Mazoyer B, Tzourio-Mazoyer N (2006) Meta-analyzing left hemisphere language areas: Phonology, semantics, and sentence processing. NeuroImage 30: 1414-1432.

Wester M (2006) Unspoken speech - speech recognition based on EEG. Master's Thesis, Universität Karlsruhe (TH), Karlsruhe, Germany.

Wu W, Gao Y, Bienenstock E, Donoghue JP, Black MJ (2006) Bayesian population decoding of motor cortical activity using a Kalman filter. Neural Comput 18: 80-118.

Yamaguchi H, Yamazaki T, Yamamoto K, Ueno S, Yamaguchi A, Ito T, Hirose S, Kamijo K, Yamanoi T, Fukuzumi S (2015a) Decoding silent speech in Japanese from single trial EEGs: Preliminary results. Journal of Computer Science and Systems Biology, 8 (5): 285-291.

Yamaguchi H, Yamazaki T, Itoh T, Hirose S, Fukuzumi S, Yamanoi T (2015b) Silent speech BCI: Extension to recognition of consonant-vowel sequences in Japanese. NEUROSCIENCE 2015, Chicago.

Yamazaki T, Yamamoto K, Kamijo K, Yamanoi T, Fukuzumi S (2012) A new single-trial EEG-based BCI using silent speech tasks. NEUROSCIENCE 2012, New Orleans.

Yamazaki T, Kamijo K, Kenmochi A, Fukuzumi S, Kiyuna T, Takaki Y, Kuroiwa Y (2000) Multiple equivalent current dipole source localization of visual event-related potentials during oddball paradigm with motor response. Brain Topogr 12: 159-175.

Yamazaki T, Sakamoto M, Takata S, Yamaguchi H, Tanaka K, Shibata T, Takayanagi H, Kamijo K, Yamanoi T (2013)

Equivalent-current-dipole-source-localization-based BCIs with motor imagery. In: Brain-Computer Interface Systems – Recent Progress and Future Prospects, Fazel-Rezai R (Ed), Pp.155-174, InTech, Croatia.

Yamazaki T, Tanaka K, Shibata T, Yamaguchi H, Ouda M, Sasaguri Y, Hirose S, Takayanagi H, Maki H, Yamanoi T, Kamijo K (2014) Categorical-data-based BCI with motor imagery using equivalent current dipole source localization. Journal of Rehabilitation Robotics 1: 1-12.

Printed by Books on Demand GmbH, Norderstedt / Germany